4.1-10.5

#781

Elementary Physics

Liquids

BLACKBIRCH®
PRESS

THOMSON
GALE

San Diego • Detroit • New York • San Francisco • Cleveland • New Haven, Conn. • Waterville, Maine • London • Munich

THOMSON

GALE

For more information, contact
The Gale Group, Inc.
27500 Drake Rd.
Farmington Hills, MI 48331-3535
Or you can visit our Internet site at http://www.gale.com

Photo Credits: **Art I Need:** 1, 2, 3, 4, 8, 10, 12, 14, 14c, 16c, 18; **The Brown Reference Group plc:** 6tc, 6tr, 6bl; **NHPA:** Stephen Dalton 16; **Science Photo Library:** Charles D. Winters 6tl.

Consultant: Don Franceschetti, Ph.D., Distinguished Service Professor, Departments of Physics and Chemistry, The University of Memphis, Memphis, Tennessee

For The Brown Reference Group plc
Text: Ben Morgan
Project Editor: Tim Harris
Picture Researcher: Helen Simm
Illustrations: Darren Awuah and Mark Walker
Designer: Alison Gardner
Design Manager: Jeni Child
Managing Editor: Bridget Giles
Production Director: Alastair Gourlay
Children's Publisher: Anne O'Daly
Editorial Director: Lindsey Lowe

LIBRARY OF CONGRESS CATALOGING-IN-PUBLICATION DATA

Morgan, Ben.
  Liquids / by Ben Morgan.
    p. cm. — (Elementary physics)
Includes bibliographical references and index.
  ISBN 1-41030-084-6 (hardback: alk. paper) — ISBN 1-41030-202-4 (paperback: alk. paper)
  1. Liquids—Juvenile literature. [1. Liquids.] I. Title. II. Series: Morgan, Ben. Elementary physics.

  QC145.24.M67 2003
  530.4'2—dc21   2003002545

Printed and bound in Singapore
10 9 8 7 6 5 4 3 2 1

Contents

Liquids splash when you pour
them or drop things into them.

What Are Liquids?

A **liquid** is something you can pour into a cup, a bathtub, or any other container. Water and orange juice are liquids. Liquids have no shape of their own. Instead, they take the shape of whatever container you put them in.

Liquids are one of the three **states of matter**. The others are **solids** and **gases**. Solids are hard objects with their own shape, like toys. You can see and touch solids. Gases are the opposite. You usually cannot see them. You cannot hold them, and they have no shape. The air around you is a type of gas.

mercury

orange juice

syrup

cooking oil

Different Liquids

The most common **liquid** is water. It fills the world's oceans. It falls from the sky as rain. And our bodies are full of it. If you look in the kitchen, you might find other types of liquid. Cooking **oil** is yellow and greasy. It is lighter than water. If you pour some into a cup of water, the oil will float on top. Syrups are made from water and sugar. They are much thicker and stickier than water.

One interesting liquid is **mercury**. Mercury is a type of **metal**. Most metals are solid, but mercury is not. It is just as shiny as other kinds of metals, though.

Snow is a form of frozen water.

Snow is good to ski on.

Freezing

If a **liquid** gets really cold, it turns
into a **solid**. When you put water
in your freezer, it **freezes** into solid
ice. Water is not the only liquid
that freezes. If you put cooking
oil or syrup into a freezer, they
will also freeze into solid lumps.

Snow is made of tiny bits of ice
called snowflakes. If you look
very closely at a snowflake, you
will see it has a star shape. Snow
does not feel as solid as a block
of ice because there is a lot of
air mixed in with it.

snowflakes

Ice melts quickly in a warm room.

Melting

Water turns into ice when it gets cold. If the ice gets warm, it **melts** back into water. Many **solid** objects melt, including iron, rock, and plastic. Ice, wax, and butter all melt, too. But they need different amounts of heat to turn them into **liquids**.

Ice only has to warm up a little bit to melt. The warmth inside your mouth is enough to melt the ice. Butter has to get much warmer than that. A hot potato will melt butter. Wax needs the heat of a flame to melt, and iron needs a furnace (a powerful fire).

The hot sun can evaporate all the
water in a desert. This makes the
ground cracked and dry.

Disappearing Water

Puddles seem to shrink and disappear on warm, sunny days. The water in a puddle, though, does not really go away. It turns into an invisible **gas** called **water vapor**. Then it drifts away in the air. In other words, it has **evaporated**. The same process makes your hair dry after a swim. It makes clothes dry as they hang on a line.

When water gets hot, it evaporates more quickly. When it **boils**, it all starts to evaporate at once. Pockets of water vapor form and burst. This makes the water bubble noisily.

Dew drops and clouds form when water vapor condenses.

Making Water Appear

Air is full of invisible **water vapor**. You can only see it after it turns back into a **liquid**. When this happens, it has **condensed**. Clouds and rain both come from water vapor that has condensed.

Water vapor condenses when it gets cold. If you take a bottle or can of a cold drink out of a refrigerator, water drops appear all over it. That is because the cold surface makes water vapor in the air condense. On cold nights, dew drops form on the ground in exactly the same way. On cold days, the water vapor in your breath condenses. It forms a cloud when your breath hits the cold air.

Insects called pond
skaters can stand on
the surface of water.

Bubbles form when
soap makes water's
skin stretchy.

Surface Tension

Fill a cup with water all the way to the top. Then, very carefully, add a tiny bit more water. You will find the water surface ends up higher than the rim of the cup. The water does not overflow because it has a kind of skin that holds it in place. The strength of this skin is called **surface tension**.

Water's surface tension is strong enough to hold small insects. If you are very careful, you can make a needle float on water by balancing it on the skin. Surface tension holds drops of water together. If you add soap to water, the skin gets stretchier and thinner, and the water forms bubbles.

Canoes float because they are full of air.

This makes them lighter than water.

Floating and Sinking

Imagine you have a handful of buttons, coins, candy, marbles, and corks. If you drop them into a bowl of water, some will float and some will sink. Things sink in water if they are heavy for their size. Coins are small and heavy, so they sink. But a ping-pong ball, which is large and light, will float. If an object is heavy for its size, it is dense. Anything that is denser than water will sink.

Ice is less dense than water, so it floats. Boats float because of their shape. The bottom of a boat might be made of dense metal, but most of the inside is air. So, overall, a boat is less dense than water.

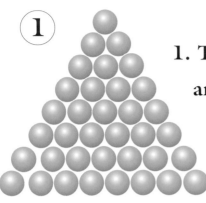

1. The molecules in a solid are packed together like bricks in a wall.

2. The molecules in a liquid can move around separately, but they always stay close together.

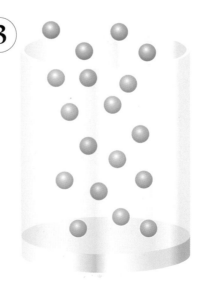

3. The molecules in a gas fly off on their own and spread out.

Atoms and Molecules

Everything in the world is made of very tiny particles called **atoms**. Usually the atoms form small groups called **molecules**. Molecules are so small that you cannot see them, even with a microsope. **Solids**, **liquids**, and **gases** are different from each other because of the way their molecules are arranged. In a solid, all the molecules are joined together strongly and held in place like bricks in a wall. In a liquid, the molecules can move around separately, but they always stay close together. In a gas, the molecules can fly off on their own and spread out.

Tie Knots in Water

Try this experiment. It shows the power of **surface tension**.

Find an old plastic container. Ask an adult to make a row of 5 holes near the bottom, about ¼ inch (0.6 cm) apart. Fill the container with water from a faucet. You should see 5 streams of water coming out of the holes. Now use your fingers to push the streams together. The surface tension makes the separate streams join into one or two large streams. You can separate them again if you brush your hands over the holes.

Glossary

atom the smallest particle of a liquid, solid, or gas.

boil turn into gas very rapidly.

condense turn from a gas into a liquid.

evaporate turn from a liquid into a gas.

freeze turn from a liquid into a solid.

gas one of the three states of matter. Gases have no shape and spread out to fill all the space.

liquid one of the three states of matter. A liquid can be poured into a container.

melt turn from a solid into a liquid.

mercury a type of metal that is normally a liquid.

metal a hard, shiny, usually solid material, such as iron or gold.

molecule a group of atoms.

oil a liquid that floats on water.

solid one of the three states of matter. Solids are hard and keep their shape.

states of matter the three forms in which all matter exists. They are solids, liquids, and gases.

surface tension a force that creates a kind of skin on water.

water vapor the gas that forms when water evaporates.

Look Further

To find out more experiments you can carry out with water, read *101 Great Science Experiments* by Neil Ardley (DK Publishing).

You can also find out more about the states of matter from the internet at this website: www.chem4kids.com/

Index